D1456969

SUGAR

Rhoda Nottridge

Illustrations by John Yates

Carolrhoda Books, Inc./Minneapolis

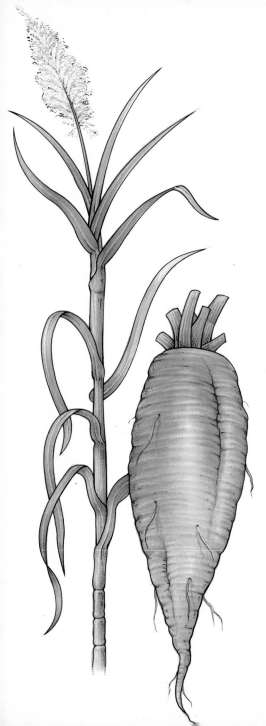

All words that appear in **bold** are
explained in the glossary on page 30.

First published in the U.S. in 1990 by
Carolrhoda Books, Inc.

Library of Congress Cataloging-in-Publication Data

Nottridge, Rhoda.
  Sugar / Rhoda Nottridge ; illustrations by John Yates.
      p.   cm. — (Foods we eat)
  Summary: Explains what sugar is, traces the history of its
cultivation, describes the role sugar plays in human diet and
health, and discusses how sugar cane and sugar beets are grown.
Includes recipes using sugar.
  ISBN 0-87614-418-0 (lib. bdg.)
  1. Sugar—Juvenile literature. [1. Sugar.] I. Yates, John,
ill. II. Title. III. Series: Foods we eat (Minneapolis, Minn.)
TP378.2.N68   1990
641.3'36—dc20   89-24028
                                                          CIP
                                                           AC

Printed in Italy by G. Canale C.S.p.A., Turin
Bound in the United States of America

1 2 3 4 5 6 7 8 9 10 99 98 97 96 95 94 93 92 91 90

Contents

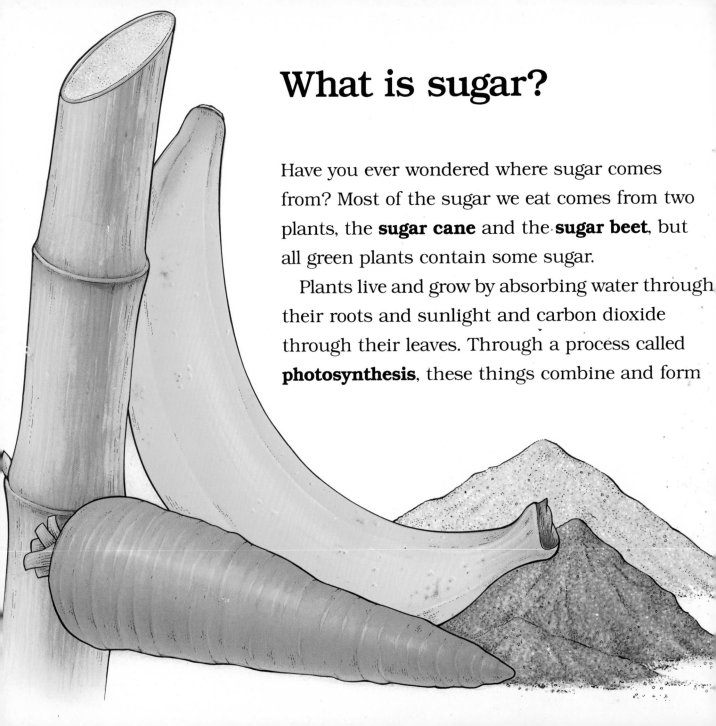

What is sugar?

Have you ever wondered where sugar comes from? Most of the sugar we eat comes from two plants, the **sugar cane** and the **sugar beet**, but all green plants contain some sugar.

Plants live and grow by absorbing water through their roots and sunlight and carbon dioxide through their leaves. Through a process called **photosynthesis**, these things combine and form a type of sugar. This sugar is the plant's supply of energy.

Sugar beets and sugar cane are put through a series of processes that extract the sugar from them. This **raw sugar** is then made into the different types of sugar that we buy in stores.

We can taste the sugar in plants when we eat fruits. Although vegetables don't taste sweet, when we eat them we are eating a plant's supply of sugar. Corn's supply of sugar can be made into **syrup**, and so can the sap of some maple trees.

Sugar around the world

There is a huge demand for sugar in the world, but the amount of sugar used in each country varies greatly. For instance, Americans use so much sugar that on average each person consumes about 95 pounds of sugar every year. In some African countries, the amount is less than 11 pounds a year for each person.

Cuba, Brazil, and India produce the most cane

This man is selling chunks of raw sugar cane in Lahore, Pakistan. They are a popular snack there.

Left: Cutting down a towering crop of sugar cane. Cane grows to a height of about 16 feet.

Below: Cane cutters on a sugar cane plantation in Cuba

sugar in the world. The USSR grows the most sugar beets, followed by France, West Germany, and the United States.

The sugar industry provides millions of jobs for people, on farms, in sugar factories, and in the transportation of sugar around the world. Some sugar companies try to help the communities where sugar crops are grown by building schools, hospitals, and other services for workers and their families.

Sugar in the past

Above: Cutting sugar cane in the West Indies in the 18th century

People have used honey as a sweetener since the earliest times, but sugar from the sugar cane plant was first used only about two thousand years ago in the Far East. Eventually, people began to use sugar from sugar cane in other countries too. It was grown in India, North Africa, and parts of the Mediterranean.

Sugar was brought to Europe from the Middle East in the early Middle Ages. European merchants

Right: Processing sugar cane at a West Indian sugar mill in 1816

8

Workers at a sugar
refinery in England
in the 19th century

soon started to buy sugar from the East to sell
in Europe. But sugar was an expensive luxury
that only the rich could afford to buy. It cost
about 70 times more than it does today.

In 1493, Christopher Columbus took sugar cane
plants from Spain and planted them in the West
Indies. They grew well in the Caribbean sun and
rich soil, so Europeans set up special farms called
plantations to grow them. The Europeans brought

people from Africa to the Caribbean as slaves and forced them to work on these plantations. The plantation owners grew rich by selling the sugar, which they called "white gold," in Europe.

In the 19th century, Britain went to war against France. The British Navy blocked off many European ports, which cut off the supply of cane sugar coming across the sea from the Caribbean plantations.

Sugar beets had been grown in Europe for a long

Making sugar from sugar beets at a French factory in the 19th century

Most sugar cane is harvested by hand, using sharp machetes.

time, but only for the food value of the beet itself. When the British cut off sugar supplies, the French discovered that they could use sugar beets to make sugar that tasted just as good as the sugar from cane. Slowly, other European countries started to grow sugar beets as well. Unlike sugar cane, sugar beets grow well in moderate climates.

Sugar cane and sugar beets are now grown in many countries. The world produces about 97 million tons of sugar every year.

Growing sugar cane

Sugar cane is a kind of tall grass that grows to a height of about 16 feet. It can grow only in a tropical climate, with plenty of rainfall and sunlight.

The sugar is stored in the **cane**, the long stalk of the plant. Cane can be harvested at any time of the year, as long as the ground is dry and the plant is big enough.

Sugar cane is usually cut down by hand, using a sharp knife called a **machete**. This is hard, slow work. In some countries, sugar cane is cut down by special machines.

When the cane is harvested, the roots of the plant are left in the ground. The following year, a new cane will sprout from the roots and grow. Additional new plants are sometimes grown from cane cuttings, also called **sets**.

Sugar cane is grown in more than 60 countries

around the world, from Australia to South America. Sugar cane provides the majority—about 55 percent—of the world's sugar supply. Most of the rest comes from sugar beets.

Unloading sugar cane for sale at a market in Brazil

Growing sugar beets

Harvesting sugar beet with a special harvesting machine

The sugar beet grows well in a temperate climate. It is grown in the United States, Canada, the USSR, China, Japan, and parts of Europe.

The sugar beet is a root vegetable, similar to parsnips and carrots. The part we eat is the root that grows underground.

The sugar beet grows in a two-year cycle. In the first year, the root builds up a supply of food in the form of sugar and swells to a weight of about two pounds. If the plant is left to grow, in its second year it will use its supply of sugar to flower and produce seeds.

Therefore, the plant is harvested at the end of its first year, when the root contains the most sugar. The harvesting machine cuts the leafy top off the beet, then lifts the root out of the ground and cleans off the dirt.

Farmers grow new sugar beet plants from seeds. A machine is used to plant the seeds in the soil in evenly spaced rows. While the crops are growing, the beet fields need to be weeded regularly. Farmers use machines to do this or spray weed killer to keep weeds from growing.

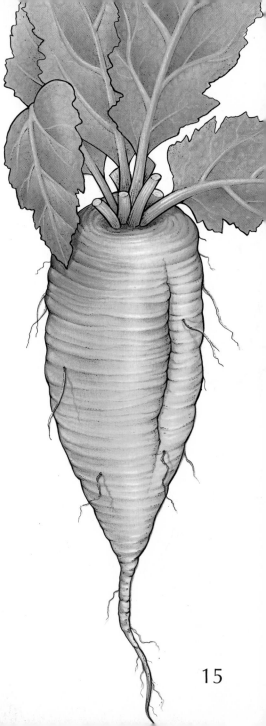

From farm to store

harvesting beets

The harvested sugar cane and beets are taken from the farm to factories where the sugar is extracted from them. Sugar beets are washed and cut by machines, and the strips of beet are soaked in hot water to dissolve the sugar.

Sugar cane stalks are usually crushed in huge machines that squeeze out the sugar. The

cane arriving at factory

16

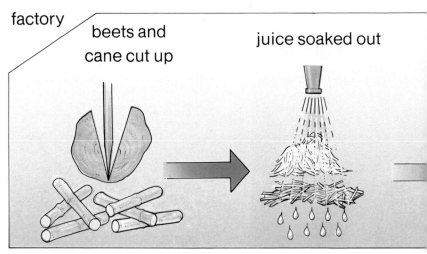
factory

beets and cane cut up

juice soaked out

crushed cane stalks are also soaked in hot water. The dissolved cane sugar is called **cane juice.**

The dissolved sugar solution is purified, then boiled until most of the water evaporates, leaving a thick syrup. This syrup is spun around in a huge drum called a **centrifuge**. Sugar crystals form and stick to the sides of the centrifuge, and the remaining syrup is drained off. The crystals are now ready to be made into the sugar that is sold in stores.

This diagram shows how raw sugar cane and sugar beets are refined to make sugar crystals.

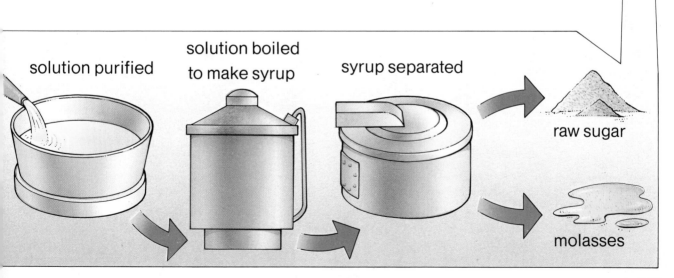

solution purified

solution boiled to make syrup

syrup separated

raw sugar

molasses

Maple syrup is delicious with pancakes. The syrup is a type of sugar made from the sap of the maple tree.

The sugar we eat

The thick, dark brown syrup that is left when th sugar has crystallized is called **molasses**. The sugar crystals are still covered in a fine layer of molasses at this stage and are dark brown in colo These crystals are known as raw sugar.

To make white sugar, all the molasses must be removed from the raw sugar crystals. The proces of removing the molasses is called **refining**.

Granulated sugar, which is often sprinkled on foods or into beverages, and **powdered sugar**, which is most often used to make frosting for baked goods, are both types of refined sugar.

Most of the **brown sugar** that is sold in stores is made by first refining the sugar and then putting back some of the molasses. The molasses adds color and flavor to the sugar. Brown sugar is often used in baking.

Above: A sugar cane juice maker in Pakistan. The cane is crushed and the sweet juice is sold to thirsty passers-by.

Right: Cookies and sugar icing make this cake a special treat for this German family.

Opposite: This machine removes some of the water from the sugar solution during processing.

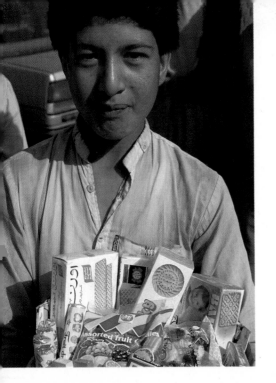

All kinds of uses

Many sweet foods contain sugar. But did you know that sugar is also added to some foods that aren't sweet? It is used in many canned and frozen vegetables, soups, cheeses, relishes, sauces, and even bread. Bakers mix sugar with yeast to make the bread rise.

Sugar is not always used as food. It is sometime

Above: A boy sells a selection of foods made with sugar at a roadside stand in Lahore, Pakistan.

Right: Sugar cane arriving at a factory on the island of Mauritius

used to preserve tobacco and leather, and even to make some paints and plastics. Some types of medicine also contain sugar. In Brazil, sugar cane is made into alcohol and can be used as a type of gasoline to power cars.

The crushed sugar cane that is left after the sugar has been extracted can be used as fuel. It can also be made into paper.

Crushing cane at a factory in Guadeloupe. When the juice has been extracted, the cane will be used as fuel.

21

Sugar and your body

Sugar is a **carbohydrate**, a type of food that gives us energy. Our bodies absorb the energy from sugar easily and use it up very quickly. For a short time after eating something sugary, we feel lively and energetic.

Eating too much sugar, however, is unhealthy. If too many sugary foods are eaten, our bodies

This diagram shows how tooth decay is caused by sugar that has turned to acid in our mouths.

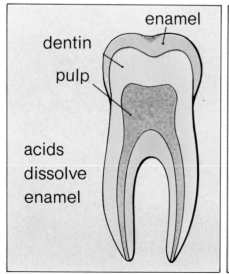

enamel
dentin
pulp
acids
dissolve
enamel

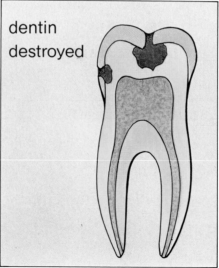

dentin
destroyed

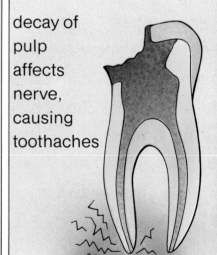

decay of
pulp
affects
nerve,
causing
toothaches

Left: It is important to brush teeth after eating sugary foods, such as this caramel apple.

Below: Sweets such as the ones pictured here are delicious, but eating too many is unhealthy.

urn the sugar into fat, and we may become over-
weight. This can lead to many serious illnesses.
Sugar is also bad for the teeth. It attacks the
enamel on teeth and helps cause tooth decay.

This does not mean that it is necessary to stop
eating sugar altogether. It is better, however, to eat
foods that are naturally sweet, such as fruit, more
often than foods that have sugar added to them,
such as cake, cookies, and ice cream.

Sugar in the kitchen

A chef makes candies from marzipan, a sweet paste made from sugar, ground almonds, and egg whites.

Sugar is often added to foods such as cereal and drinks such as tea and coffee because we like th sweet taste. It is also used in making cakes, cookie and other desserts.

Sugar acts as a **preservative** that helps keep mold from growing on food. It is an important ingredient in jams and jellies.

Beautiful decorations for food can be made using powdered sugar. The sugar is mixed with a liquid and sometimes with food coloring to mak a smooth icing. This icing can be made into mar shapes. As it dries, the powdered sugar hardens. The picture on the right shows some ways of usin powdered sugar icing.

Try decorating some cookies with your own designs, using colored icing and sugar. You can make patterns, faces, or whatever designs you lik

Honeycomb

You will need:

1 tablespoon butter or margarine
1½ cups brown sugar
6 tablespoons honey
1 teaspoon vinegar
4 tablespoons water
2 teaspoons baking soda

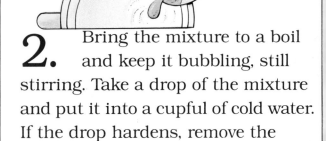

2. Bring the mixture to a boil and keep it bubbling, still stirring. Take a drop of the mixture and put it into a cupful of cold water. If the drop hardens, remove the saucepan from the heat.

1. Put the butter or margarine, sugar, honey, vinegar, and water into a saucepan. Cook over low heat, stirring constantly, until the sugar has completely dissolved.

3. Stir in the baking soda and pour the mixture into a small baking pan that has been greased with butter or margarine.

4. Let the honeycomb cool, then break it into pieces.

Coconut squares

You will need:
8 tablespoons condensed milk
1½ cups powdered sugar
¾ cup shredded coconut
a few drops of pink food coloring
1 tablespoon cocoa powder

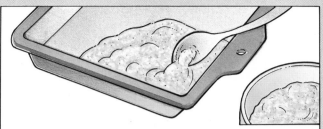

2. Divide this mixture in half. Add the food coloring to one half and spread it into the cake pan.

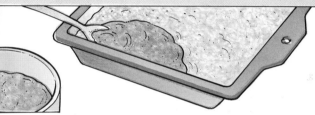

3. Sift the cocoa powder into the other half of the mixture and stir well. Spread this on top of the pink mixture in the cake pan and let it harden.

1. Lightly grease a square cake pan. Mix the milk and the powdered sugar. Then stir in the coconut.

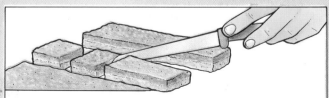

4. When the mixture has hardened, cut it into squares.

Chocolate chip cookies

You will need:

6 tablespoons butter
6 tablespoons brown sugar
1 egg
½ teaspoon vanilla extract
½ cup plus 2 tablespoons flour
¼ teaspoon salt
6 ounces semisweet chocolate chips

2. Add the egg and vanilla extract and mix well.

1. Preheat the oven to 350°. Put the butter and sugar into a bowl and mix well.

3. Stir the flour and salt into the mixture. Mix well.

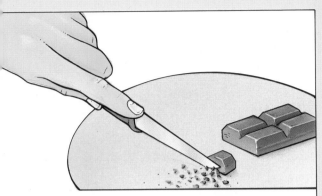

4. Gradually mix the semisweet chocolate chips into the cookie dough.

5. Drop the cookie dough by teaspoonfuls onto a cookie sheet, leaving some space around each cookie. The cookies will spread out while they are baking, so make sure that they are not too close to each other.

6. Put the cookies in the oven and bake them for 8 to 12 minutes. Be very careful when using the oven and be sure to use oven mitts or potholders.

Glossary

brown sugar: refined sugar that has had molasses put back into it

cane: the long stalk of the sugar cane plant, in which the plant's supply of sugar is stored

cane juice: the sugar solution that is produced when sugar cane is crushed and soaked in hot water

carbohydrate: one of a group of foods that provides energy

centrifuge: a machine used in sugar processing. A dissolved sugar solution is spun rapidly in the centrifuge, causing sugar crystals to form.

granulated sugar: coarse-grained sugar that is the most commonly used form of sugar

machete: a long, sharp knife, often used to cut down sugar cane

molasses: the thick syrup that is left when the water is boiled out of a sugar solution

photosynthesis: the process by which plants form the sugar that is their supply of energy

plantations: farms that are specially set up for growing one type of crop

powdered sugar: a form of refined sugar that is often used in baking and can be mixed with a liquid to make icing for cakes and cookies

preservative: a substance that helps keep foods from rotting

raw sugar: sugar crystals that have not been refined

refining: the process of removing molasses from sugar crystals to produce white sugar

sets: plant cuttings from which new plants are grown

sugar beet: a root vegetable that is grown in many countries for its sugar

sugar cane: a type of grass that is grown for its sugar and provides much of the world's supply of sugar

syrup: a thick, sugary liquid that is used as a sweetener

Index

Photo acknowledgments

The photographs in this book were provided by: pp. 6, 19 (top), 20, David Cumming;
pp. 7 (top), 14, 23 (bottom), British Sugar Bureau; pp. 7 (bottom), 19 (bottom), 20 (bottom),
23 (top), 24, ZEFA; pp. 8, 9, 10, Mary Evans Picture Library; pp. 11, 13, 21, Hutchison
Library; p. 18, Tate & Lyle.